# THE DESPERATE DELTA

## Social Ecology of Sunderbans

# THE DESPERATE DELTA

## Social Ecology of Sunderbans

KUMAR CHATTOPADHYAY

GYAN PUBLISHING HOUSE

New Delhi 110 002

THE DESPERATE DELTA
Social Ecology of Sunderbans
*Anthropology, Geography, Ecology, Environment*

ISBN : 81-212-0484-4



Published in India by
Gyan Publishing House
5, Ansari Road
New Delhi 110 002



Printed at Mehra Offset Works, Delhi-110 002

# Contents

# 1

# The Backdrop

*Praying Mantis* is a patient insect. For quite sometime it watched the dragonfly to settle on the branch. It started stalking so slowly and stealthily that not a leaf moved, not a twig twisted. Moments later, the main dish was over. The mantis then methodically went on finishing the bits and morsels which were once a dragonfly. Hunger being satiated it relaxed and thereby committed the greatest and incurable blunder of its life. Relaxation is forbidden in this part of the world—the Sunderbans. The mantis learnt it too late.

Far from the madding crowd, beyond the jungles of concrete of the city and towns, where the aquamarine waters of the Bay of Bengal embraces the emerald greens of the forest, twice daily, lies the Sunderbans—the largest delta of the world. Geographically it is located between 21° 31' and 22° 38' North and 88° 5' and 90° 28' East.

"Sunderbans" literally means beautiful forest. Beautiful, indeed, it is. But as to the origin of the name, there are two acceptable hypotheses. One is that at one time the dominant tree of the area was `Sundri` (Heretiera litoralis) and hence the name. The other is the nearness of sea, `Samunder` in local dialect. So a forest or `ban` near the sea or `Samunder` becomes `Samunderban` or the Sunderbans.

By the Geologist's scale of land formation, the development of this coastal area is rather a recent phenomenon. The alluvial condition of the area is hardly conducive to the fossilization of animals and plants. Naturally, one has to search for other evidence in `digging up its past`.

Unfortunately, the nature of the surrounding terrain gave little scope for wider archaeological excavations. Some surface findings of artifacts scattered in various places in and around the Sunderbans are the only clues. The earliest among them, a polished hammer and pestle, obtained at Harinarayanpur beyond Diamond Harbour, dates back to the neolithic culture akin to that of Tamralipta in Midnapore district.

Compared to the prehistoric evidences, the historic records of the Sunderbans are relatively little better though not exhaustive. We learn that, "Five hundred years ago, one Khan Jahn possessed a 'jagir' in the Sunderbans up to 1459. A century later, one Daud obtained a 'raj' in the Sunderbans followed by his son Pratapaditya, who was later subdued by Man Singh of the Moghul army." For about two successive centuries that followed, the

Sunderbans did not figure much in the records. At the most, it appeared only as a water-wash in the easel of the contemporary historians' painting of the country.

Meanwhile, the Ganga changed its main course from the Bhagirathi to the Padma leading to concomitant ecological changes and hardship to its inhabitants. As if this natural calamity was not enough, the Portuguese and Magh pirates swooped down time and again, plundered, ravaged and tortured the people of this delta to no limit. It has been reported that "according to the diary dated 30 November 1676 of Streynsham Master, they had carried off the people of the river-side villages to their slave market at Pipli about 1664" (Das)..

There is no knowing if the salinity of the waters increased, as more and more blood and tears thus poured into the rivers. What however is certain is that the area gradually became desolate and depopulated. Tortured and neglected, the delta since then has been constantly trying to get a 'place under the sun' against all odds.

If you keep your eyes open while meandering through the creeks and canals of the delta, you are likely to find pockets of ruined human settlements. You may be tempted to call them renants of civilization, but I do not. I identify civilization as an all round development of resources, both natural and human. And that too as complementary to each other and not the development of one at the cost of the other. But the Sunderbans had a 'fait accompli'. The petty merchants of the other day became, by conspiracy, the rulers whose rights there was none to dispute.

The East India Company in 1765 acquired the 'diwani' of the Sunderbans. Till that time the green world of the delta hardly experienced the sharpness of the lumberman's axe. Mr. Henckel, from whom the present day Hingalganj derived its name, was the first to attempt in 1782 to clear a portion of the forest to lease it out for cultivation. The purpose was to gain revenue from the so-called unproductive land and to obtain a store of rice against seasons of drought.

The days of sorrow, however, continued unabated. The dacoity on boat-routes, illicit tolls levied by the adjacent zemindars, oppressions by the 'malgis' on the 'mahindars' or salt-boilers rendered the scheme practically defunct. The attempt of Henckel to bring 7000 acres of the Sunderbans under cultivation, however, paved the way for subsequent human settlement in the area.

During the 19th century a number of different surveys on the Sunderbans was made, one Sunderban Commissioner was appointed, several Rules were promulgated and also postponed, on the leases of lands. For instance, according to 1879 Rules, the grant of land was made to two classes of people: (a) blocks of 200 acres or more, leased for 40 years to large capitalists for development of the area by spending money, and (b) blocks with less than 200 acres leased to small capitalists for clearance by the cultivators. The system led to the growth of an undesirable class of land speculators and middle--men, and to the grinding down of the actual cultivators by excessive rents. In 1804, large Capitalist Rules were abandoned and a 'Rayatwari' system was introduced. In 1905, the post of Sunderban Commissioner was abolished, and the Commissioner's power and function were vested in the revenue collectors of the district. In 1909, the 'Rayatwari' scheme was also abolished and the new system of 'Rayatwari' settlement or colonization was introduced in 1919 by virtue of which *raiyat* had been settled on the estate.

Attempts in the line of Mr. Henckel was also made later by Sir Daniel Hamilton at Gosaba which was once called Hamiltonganj, after Hamilton. Perhaps, he was the first man to include rural welfare and cooperative approach under the scheme of reclamation and spatial development in this part of the country. No less a person than Rabindranath Tagore himself visited the place and appreciated the venture.

According to the district gazetteer (1914), farmers in the Sunderbans seasonally migrated from outside. They engaged themselves in cultivation at their homes during April-June, then used to come to 'Abad' in July-September, returned home-land and harvested in 'November-December' and then back again to 'Abad' to harvest in 'January-February'. This was their annual cycle of allocation of seasons to successive stages of agricultural work.

Besides cultivators, wood-cutters would come from Backergunge, Khulna, Faridpur, Calcutta and other places for jungle-clearance, earth-work, etc. during October and May. Labourer for this purpose used to come from Hazaribagh, Birbhum, Manbhum, Bankura and such other places. In the remote part of the Sunderbans, however, 'husbandsmen' of the land dwelled permanently. Since these cultivators and other labourers could not move in or out easily, some of them began to live there out of necessity for longer periods and eventually settled down permanently. Over the generations such settlements gradually developed into villages of today.

Obviously, in such areas, during the initial phases of uncertainty and even high risk of life, people belonging to some kind of a closely knit social group stayed together to work jointly and help each other at times of need. Thus people of identical kingroup or community might have started to settle in the same area. People at the lower rung of the society—low castes and tribals—usually lack in economic security in the wider world and often float like nomads in search of opportunities to survive somehow and settle down. Mostly streams of such people started flowing in and crowding around the Sunderbans. The leaseholders also employed various types of employees, guards, labourers, etc. who belonged to such low castes and tribes and many of them had to work and stay in the Sunderbans. People of lower social ranks thus continued to migrate more than the others from the early generation. The increase in population in subsequent generations further strengthened this pattern of distribution of different low castes and kingroups.

Following Allan and others, it was found that low-caste Hindus and Muslims resided in the north of the Sunderbans. Among low-caste Hindus, 'Pods' were numerous in the west, 'Namasudras' and 'Chandals' concentrated in the east; and among the Muslims, the sect 'Fargis' was dominant. This was then the settlement morphology of the Sunderbans during the early British rule. And it was apparent that their scheme of development always looked for reclamation of forest and introduction of land for cultivation. Concomitant to such approach the hydraheaded destruction of natural resources took its birth. The irony of the situation is that while this is perhaps the richest area in forest resources in West Bengal the wealth is not being properly utilised either because of ignorance or due to the greed and avarice of a handful of men who exploit them for their personal aggrandisement. In any case, even today, planned utilization of such resources for the society *per se* is sadly missing.

# 2

# The Stages

This delta is breathing under the over-lapping influence of three major ecosystems. These are the forestry, the water ways and the human settlements. Their exclusiveness apart, marked by respective physical structure and spatial distribution of resources, the symbiosis is reflected functionally through the exploitation of resources mainly by the human groups.

The uniqueness of the Sunderbans is well exhibited by its flora and fauna. Contextually, the major share of such wealth is restricted within an area of 2585 sq. km. of forest. This has been declared as a National Park and recently as a Biosphere reserve. The total area is divided broadly into two zones: the core area and the buffer zone. The core area is of 1330 sq. km. covering Mayadwip, Chhoto Hardi, Gosaba, Matla, Gona, Chamta and Bagna blocks. Of these, the Chamta blocks with 124 sq. km. are also termed as the primitive area. The rest 1255 sq. km. are known as buffer zone.

With the passage of time and human carelessness, the forest has lost much of its faunal treasure. Animals like Rhinoceros, Buffaloes etc., have already become extinct. At present, important animals are tigers, chital or spotted deer, bears, different types of cat (Felis sp.) birds of different varieties, both local and migrants, and varieties of snakes. It is tragic to note that although snake-bites take a heavy toll of lives in this area, no attempt has been made to prepare anti-venom injection by engaging local people. This could also have generated some employment, apart from saving the poor lives.

Some of the dominant floral wealth of the area are 'sundri', 'keora' 'goran', 'golapta', etc. The trees are good sources of forest produce in the Sunderbans. Not much work, however, has been done to *estimate* the total resources of this littoral and swampy forest which is rather rare in other part of the world. Unfortunately, no comprehensive lists of botanical species, based on scientific inventory and empirical research with the representative coverage of the total area, are readily available. We have only a partial knowledge and information on the dynamics of the Sunderbans forestry. Our enquiry, which is also very limited, on the above issue, has revealed the following facts.

The ever-green forests of the Sunderbans as a type belong to the mangrove class. There is, however, a distinct spatial-structural distribution of species conglomeration. Near the water line, lands are inundated twice due to tides and *dhani ghas* (oryza coarctata) grows in the first level followed by the pneumetaphores on the next. Other smaller *root growing* trees sequentially follow and are functionally correlated with the height of the land from the coast line and make way finally to the big trees (wherever these may grow) at the upper-most level.

Since this deltaic region does not rise sharp from the water line, the general floral pattern appears to be islands covered with ever-green and dense undergrowth of shrubs fenced by pneumetaphores, with widely spaced big trees providing the canopy. Contextually, the following exercise is intended to provide an idea on the interaction of the two ecosystems—the floral and human—through the process of functional utilisation of one by the other. *Gol pata* is utilised for covering the roof of the huts; the *Geela creeper* substitutes for jute rope while *Hental* leaves are used for manufacturing sleeping mats. The *Bines, 'Tamarinds, Bablas* are used for preparing plough, while *Dhondool, Bine, Halud* etc., are needed for constructing yoke for the bullocks. For manufacturing boats and as posts and beams the *Sundaris, Keòras, Pasure* are very much favoured. The *Geowa* trees are utilised for making packing-boxes, match-boxes as also for manufacturing match-sticks. The *Dhundools* are also in great demand in the furniture-making circles.

The above list of trees is by no means an exhaustive one. It is a fact that we have only a limited idea of our renewable resources in the *Sunderbans,* most of which are being systematically plundered commercially.

Proper assessment of our national wealth in this region is an imperative need which is long overdue. Collaboration of a team of biosocial scientists, naturalists and statisticians can only solve the issue whose importance from the point of engulfing pollution, economic necessity, ecological balance and wealth can hardly be overemphasized.

Even courting the risk of being repetitive, it must be stated that hardly any systematic effort has been made by the concerned government organizations for an inventory of these natural wealths. And whatever limited data we could muster was possible by our own efforts only and through the help of several individuals in their personal capacity. Sadly, none has tried so far to *estimate* the *number* and *composition* of species; whether they are increasing or decreasing and the economic viability of their largescale use. Any knowledgeable person, we hope, will readily admit the necessity of such estimates which will throw light on the potentialities of development or otherwise of our natural wealth.

The single natural phenomenon that played and has been still playing a vital role during the genesis, gesticulation and development of this delta, is the ecosystem of the waterways.

Water is synonymous with life. The epicentres of all the great civilizations were built and flourished by the river sides. Such a great civilization like the Indus Valley developed through river communication system. Exchange of goods between countries was facilitated via waterways.

This life-sustaining river system, however, was not always a boon for all, at least, not for the Sunderbans for centuries in the past. This reverine ecosystem which covers an area of 1780 sq. km. often cuts into the forests and has bounded the Sunderbans of 9629.9 sq.km. from the three sides. On the west is the river Hooghly, on the east is Raimangal which, incidentally, marks the international border with Bangladesh; while, the ultra-marine waves of the Bay of Bengal lap the south. Between Hooghly and Raimangal there are several sub-systems of rivers. Some such rivers are: Matla, Thakuran, Baidya, Saptamukhi, Gosaba, Herobhanga, etc. These rivers with their tributaries have criss-crossed before their confluence with the Bay of Bengal to such an extent that it has created 55 islands in the area.

While propelling through the creeks and canals you are sure to note that most of the villages lie below the heights of the river beds. This resulted, time and often, in over-flooding of the villages and the problem of drainage so much so that during monsoon the whole area turns into a quagmire of slippery mud and water. To add to their sorrow, the seasonal cyclonic calamities often ravage the human settlements all around.

These rivers, at certain points of history, tempted the river pirates to thrive. According to the *East India Chronicle*, the Maghs pirate carried off 1800 men, women and children to Arakan in 1717 and sold them as slaves (Das).

The piracy has not halted even now. There has been a change in the method only. Instead of physical captures, the modern pirates exploit the poverty. They occasionally allure the girls and children who ultimately find themselves as prostitutes and virtual slaves in some neighbouring or middle-east countries. Seldom, the uplanders care for their losses or try to wipe out their tears.

Dear reader, what could be the saturation of salinity of the Bay of Bengal? How many litres of human blood and tears Ganga and Padma are needed to turn them into a sea and spell a curse of perpetual thirst of Tantalus to the indifferent people of the foreland?

# 3

# The Actors

This fairly big eco-stage of the Sunderbans holds a sizeable number of actors. Some of them are struggling for survival while others are engaged in a continuous process of subsistence in extracting their livelihood. There is no denying that different types of skills are necessary for pursuing different jobs. This is more true in a deltaic forest area. It presupposes a conglomeration of people of certain segments at certain points of time. They either had traditional expertise or were adaptable to the forest-based operations. Let us look into this matter.

According to the 1981 census, the combined percentage of the scheduled castes and the scheduled tribes in the state was 27.62. Such percentage was 26.80 in the district of 24-Parganas. But, in the Sunderbans, which is within the district of 24-Parganas, noticeable variations are observed. For example, in the 15 Police Stations that demarcate the Sunderbans, the combined average percentage of the scheduled groups was 42.78. But in certain police stations such percentages exceeded even 70.

These figures reveal high concentration of the scheduled communities, and consequently, of lower class of people, in certain pockets of the Sunderbans. Apropos, at certain point of time, this area has provided livelihood to some of the backward groups within its deltaic forest fold. They were the first settlers to exploit the forest resources. And these groups have their own social history behind such habitation.

However, over the centuries, this area has not changed much from a sort of more or less monocrop rice cultivating zone. The total land is about 15.23 lakh hectares of which about 4.3 lakh hectares were cultivated in the previous year. Although the average yield of the 'Aman' paddy in the district is not much less in comparison to the average yield of the state *per se*, the severe droughts concurrently for two years (1981-82) have pushed the district into near famine condition. And several cases of death due to starvation have been reported from several Police Stations in the district. Besides, due to the climatic catastrophe the 'Ravi' crops, to whatever extent it had started, vegetables, and even the cultivation of water-melons have suffered immensely.

In West Bengal, the Sunderbans today have an area of 9630 sq. km. with a population of 24 lakhs. Although 95 percent of the people in this rice-cultivating zone depend on agriculture, more than 54 percent are landless cultivators. Even more than 85 percent of those who have lands can be classified as marginal farmers. They are one of our actor groups fighting the crisis of subsistence. Most of them are almost perpetually in debt. An enquiry shows that more than 50 percent of such families are indebted and on an average the amount of such debt per family is Rs. 787.18 p.

According to the 1981 census, percentages of workers to total population in West Bengal and 24-Parganas district, as a whole, were 30.18 percent and 26.10 percent while in the part of 24 Parganas under study the same was 25.77 percent. Among the workers in "forest division" of the district, 77.42 percent was engaged in livestock, forestry, fishing, hunting, plantation and allied activities. On the other hand, only 5.0 percent were cultivators and 3.55 percent were agricultural labourers. Thus, the people of the forests of 24-Parganas, mostly depend upon the forest and forest products for their livelihood. It may also be remembered that a substantial proportion of such people were and are engaged in exploring the natural resources through fishing and gathering. That is why the reporting of this fantastically high percentage of *workers*, in the context of this region, is not surprising. It only suggests that each one has got to go out to earn whatever is available.

Associated with the ecosystem of this delta, there are five major 'actor groups'. Two of them, the cultivators of land (agriculturists) and captive breeders of fish (bheri owners), are engaged in maintaining their respective levels of living. While the former are toiling at the level of subsistence, the latter are mostly accruing enormous margin of profits.

I believe that every paisa gained by means other than one's own labour carries behind it a history of exploitation. I also believe that such exploitative intentions are the *niche* of homosapiens and are more or less universal.

Of the three remaining groups, collectors of fish, honey and timber are facing the challenge of survival.

It is a fact that in the monocrop area with high salinity, the peasants have to labour hard. Nevertheless, they have this resource of land to clutch or fall back on during any calamity.

The word 'calamity' is not to be found in the dictionary of the rich *bheri* or fishery owners. Their resources are their ponds and the fish spawns supplied, practically throughout the year, by small children, the olds, and others. For a full day's toiling, they are paid miserly by the *dalals* or go between who collect these spawns on behalf of the bheriwallas. Prices, again, vary widely and seasonally. Accidents to the catcher of spawns do occur sometime. And for an average income of Rs. 3 to 4 per day, a score of them have lost one or the other of their limbs under the jaws of *kamots* (shark). Their blood only adds salinity to the water and does not move the bheriwallas to shed even crocodiles tears. These spawn collectors have nothing to lean to during necessity. Their skill is the only resource to earn a living and also to save them from calamity.

Unfortunately, these spawn collectors, on the other hand, are also depleting the natural resources of this delta. While selecting their spawns, from their nylon dragnets, they throw away other frys. Throwing of such frys by about 200 to 300 collectors everyday indicates the enormous wastage of aquatic resources. If we wish to save our future aquatic resources from complete destruction, these spawn collectors should be taught to release other frys back into the water. They must be helped in utilising their skill in picking up their own requirements and releasing but not destroying the other species.

The fishermen and honey collectors of the rivers and forests of the Sunderbans are our other groups of actors struggling for their survival. We carried out our investigation among them. Certain details are analysed below.

It should be stated at the outset that to carry out an empirical research in these areas, is a formidable challenge. These areas remain for half of the year practically cut off from the main land due to rains and storms which are very frequent in this deltaic part of West Bengal. Even beyond rainy seasons communications only exist through waterways. As a result, it takes considerable time to establish such communication as the local waterways are tidal channels. This means that the journey time of a boat is dependent on the ebb-tide cycle of the river.

Furthermore, at the time of our pilot-enquiry there was no readily available information which could help us in locating the people or villages that utilised the forest products for their benefits, economic or otherwise. However, a chance encounter with certain individuals helped us in locating an organisation from which we collected some data. Analysis of these data proves, that in view of non-availability of relevant information these can be and should be utilised for getting, at least, a birds eye view of the problem of human adjustment.

Data were collected from 4 field stations under the jurisdiction of directorate of Project Tiger of the Government of West Bengal. Collected information was related to the permit holders who were allowed by these stations to enter the jungles of the Sunderbans for collecting honey, timber and fish. According to the latest information, this district has 6 sub-divisions, 51 police stations with 3792 villages. On the other hand, the area known as the Sunderbans within the district consists of 15 Police Stations having 1375 villages.

Contextually, if an association between the proximity of villages from the forest and exploitation of its resources is assumed, we can expect 36 percent of the villages of this district to derive such benefits since these villages are in the same area.

Against this backdrop, let us examine our data obtained through pilot enquiries during 1982-83. The first information provided by our data was that a total of 6483 groups of permit holders visited forests for different purposes. They came from 157 villages under 19 Police Stations. It was observed that 3 out of 15 Police Stations of the Sunderbans were not connected with forest exploitation while 7 Police Stations, which were beyond the ambit of the Sunderbans, were involved in such operations. These data, only apparently point towards the common expectation of linkage between proximity and exploitation but nothing beyond.

The number of groups of permit holders coming from each of these village ranged from 1 to 1154 corresponding frequency of villages varying from 36 to 1. The distribution apparently indicates an inverse relation between the groups and the villages. In other words, when the number of groups increases, the number of villages they come from decreases and *vice versa*. Moreover, 19 villages out of 157, i.e. , 12 percent, were sending 75 percent, of groups. Do these data indicate a geographical or spatial concentration of forest exploitation activities by such groups of villagers who were within an accessible zone?

This leads us to the question of identification of such areas or zones, if possible. First, dichotomy may be constructed by the areas falling either within the Sunderbans or outside. In view of the above, we computed the average number of groups per village from which these groups had come. Also, the percentage of such villages in a Police Station to total number of villages within it was computed. It was observed that while no villager approached for forest permits in 3 of the Police Stations of the Sunderbans, villagers in 7 Police Stations, which are beyond the boundary of the Sunderbans, did participated in collecting forest products. Interestingly, we did not find a single case of permit holder coming from places beyond the district of 24-Parganas. Moreover, using the two sets of percentages just mentioned statistical tests of contiguity ratio* point to the existence of contiguous areas or zones with differential permeability of the forest exploiting activities by the people of the district. Arranging the data in a map and then testing for contiguity we get, instead of a dichotomy, three identifiable zones consisting of 74, 32 and 51 villages in 5, 4 and 10 Police Stations respectively. These three zones will be labelled, on the basis of their proximity to core forest area as nearest (Zone I), nearer (Zone II) and distant (Zone III) zones respectively.

One interesting point is that 3 of the Police Stations of the Sunderbans show more closeness in this context with 7 Police Stations that are beyond the Sunderbans. We also noted earlier the wide range of variation of the number of groups from one village to another. This raises the question whether such dispersion suggests any extreme characteristic and may be, therefore, statistically neglected. Statistical test and inference on this issue, however, relate that contrary to the above suspicion such range and values probably reflect the inherent character of the universe and should be taken into account. Statistical test on the relation between permit holders of villages and their distribution by zones shows somewhat complex characteristics. For instance, while the comparative values

---

*Geary, R.C.: The Contiguity ratio and statistical mapping. *The Incorporated Statisician*, November 1954, pp. 115-145.

of mean indicate a ranking pattern of zones, the high values of s.d. possibly contradict such an inference. However, when the data showing the distribution of the villages sending permit holders to the forest are plotted on the physical map of the district, a spatial clustering or zonal pattern emerges.

The idea behind the data that are presented next is to probe whether there is any relation between the zones and number of groups representing varying number of villages. It appears that, considering the large-sized groups i.e., 100+, the zones are playing differential roles of influence. For example, out of 19 such groups 14 belonged to Zone I followed by 3 and 2 in Zones II and III respectively. Data appear to substantiate, to some extent, the importance of proximity in utilizing forest resources by a relatively large number of people.

A question might, however, be raised at this stage whether it was the zones or the constituent villages that played the vital role in this regard. In other words, whether particular village(s) had specific characteristics to influence participation by such large groups etc. But before analysing the data within the Zone I, we isolated Gosaba P.S. from the cluster because it is nearest to the forest and is also separated from the rest by natural waterways. This was done to avoid undue influence by one P.S.

Results of the test of significance of Kolmogorov Smiranov's D-Statistic, however, prove that there is no intra-zonal variation. And the construction of these zones earlier by plotting the primary data on the relief map was quite logical and statistically tenable.

Till now, we have established that (a) there is an inverse relation between the number of groups entering the Sunderbans from certain number of villages and (b) there is a definite pattern of zonal differentiation in terms of participation of groups in the activities of collecting forest resources. Furthermore, the statistical tests done so far also prove that the occurrences of large groups of permit holders in different zones are not merely by chance but possibly reflect habitat characteristics.

We next utilised the data to probe into the relations between the participants (groups) and their purposes for visiting the forests of the Sunderbans. The distribution showed that out of the total 6483 groups, 81 percent, 9 percent and 10 percent came for fishing, collecting honey and gathering timber respectively. This was the overall pattern. The question that cropped up was whether such pattern was also followed in the zones. Consequently, we found that exclusively for *fishing* 72 percent, 19 percent and 9 percent of the groups came from the Zones I, II and III respectively. Such percentage of groups engaged in collecting *honey* was 98 in Zone I, the rest came from Zone II while none from the Zone III. In respect of groups connected with *timber* collecting activities, a zonal difference was also noticeable. Maximum participation came from the Zone I followed by Zone III and Zone II in the said order according to proportional representation.

Above data highlight the following points:

(a) irrespective of purposes, Zone I was sending maximum number of groups to the forests, (b) collecting of honey seemed to be a monopoly activity of Zone I and (c) Zone III appeared to be comparatively more involved with fishing and collection of timber.

Data on the number of villages per zone sending out groups to the forests for different purposes showed a different order of ranking of the zones. For example, it was the villages of Zone I that seemed to be maximally involved in forest operations, followed by villages from Zone III and Zone II respectively. These evidences of involvement of villages of Zone I, sending maximum number of groups to the forests, probably indicated formation of *niches* by certain communities living within a specific spatio-ecological zones. We propose to examine this question later.

We examined our data by categorising the purposes of the groups in terms of their exclusive involvement in any of the three activities of fishing, honey collecting and timber gathering. Subsequent probing was on the distribution of villages, by zones, incorporating the combination of purposes. Major features that emerged from these data were (a) collection of honey, with or without any other activity, restricted to Zone I and involving 24 percent of the total villages; (b) predominantly fishing activity followed by 34 percent of villages seemed to cut across the zonal boundaries while (c) gathering of timber was mostly controlled by the villages of Zone III.

We next tried to examine whether (i) the average number of permit holders of different Police Stations and (ii) the percentage of villages, under each of such Police Stations, indicated a pattern of concentration, in terms of different purposes of visit to the forests, by the concerned villagers.

Data showed that the size of the group fishing in the forests varied from 1 to 106. The corresponding size in the case of collection of honey and timber ranged between 1 to 20 and 1 to 27 respectively. Again, percentages of the involved villages to the total by Police Station for groups visiting the forests for fishing, collection of honey and timber ranged between 1.15 to 29.03, 0.73 to 29.03 and between 0.70 to 18.39 respectively.

Statistical test of contiguity ratio when applied to these data showed that on both the counts, the activity connected with honey indicates clustering. Timber collecting operation, on the other hand, showed scattering both in terms of average size of the group and percentage of villages involved. Fishing, however, played a somewhat intermediate role. While it indicated a certain tendency of forming cluster in so far as the proportion of villages involved in sending the groups is concerned, it did not show any pattern of contiguity or concentration by virtue of their average sizes.

The above inference was further strengthened by subjecting the data to the Kolmogorov-Smiranov test. It was done by grouping the villages sending permit holders, for the three purposes (neglecting the combinations there of) of fishing, honey-collection and timber gathering. We found that collection of honey as occupation significantly differed from those of fishing and timber-collection. Again fishing as occupation also differed from that of timber-collection significantly. In other words, the data probably indicated the specific importance of each of these activities. Such data possibly marked, although by different degrees, a pattern of continuous distribution.

So far we have examined our data in terms of spatial distribution by groups, villages, and purposes of visit by these groups. We also noted the zonal distribution of the villages, by Police Stations, that were sending these groups or permit holders to the forests. We now propose to examine whether the Police Stations containing these villages of permit holders had certain specific characteristics that might have influenced these groups to adopt such forest-exploiting activities.

Theoretically speaking, we could have selected any number of characteristics of the Police Stations, available as secondary data in the census, and examine the same against the above premis. We, however, selected in all, 11 characteristics of the Police Station. Of these, two were demographic, six economic and the remaining three reflected community qualities.

Under the demographic characteristics we took into account the total male population and the density of population. The selection was influenced by the fact that all the permit holders were male persons, albeit for obvious reason. Further, an area with high density of population may have to oblige certain segments of population to take to jobs in different areas for their survival and also to minimise the local competition.

Similarly, villages having much unemployment and unsettled or semisettled labour force in the agriculture sector etc. under economic constraints might have played a role in sending these permit holders to the forest. Much of these spatial characteristics possibly indicated the saturation of the carrying capacity of the area. On the other hand, it may be that the culture-specific of certain community and\or their backwardness had a certain role to play vis-a-vis these forest operations.

These data, when tested statistically, produced interesting results. The values of Spearman's rank Correlation Coefficient show that out of the 11 characteristics 4 show negative correlations. This means that the total male population, density, degree of unemployment and the total labour force in agriculture, (more or less seasonal in nature) not only had no correlation but played reverse roles. That is, they are not influencing the permit holders to venture on work in the forests. Of the remaining 7 characteristics, high correlations have been obtained between the community and permit holders.

In other words, this test indicated a possible occupational specialization by the Scheduled Community in the exploitation of forest resources.

Taking a cue from the above let us now examine the proportion of the Scheduled Community of permit holders who entered the forests.

Contextually, we find that for Zone I the average percentage of Scheduled Caste male population ranged from 48 to 72 with the average of averages being 62.60. In comparison, Zone II showed a range between 23 to 65 while such range for Zone III was from 13 to 63. The average of averages for these two zones were 43.20 and 31.39 respectively. The data seemed to point to an ordering in the distribution of the Scheduled Castes by zones. This might have played a role also in the formation of the groups entering the forests. In other words, it may be that the zones having predominance of Scheduled Castes were sending large number of them and thereby indicated, indirectly of course, a possible association between the Scheduled Castes and exploitation of forest resources. Did it not also indicate the formation of *niches* by these groups?

We now propose to examine our data on the group level. That is, whether the villages that were sending large number of permit holders also have Scheduled Castes as dominant group. However, we will consider as illustration, only those villages which sent larger groups. We already found that altogether 19 large groups, in terms of number of groups being 100+, from 19 villages and 7 Police Stations, participated in forest-going activities. Of them, 14 villages belonged to Zone I, 3 to Zone II and 2 to Zone III. If we classify these data by the Sunderbans and non-Sunderbans area then, except 2 villages of Bhangar and Sonarpur Police Stations, sending 114 and 176 groups respectively, all the other villages were within the Sunderbans area.

Interestingly, proportional distribution of the Scheduled Caste population in these villages indicated that those two villages, which were sending large number of groups and which were outside the Sunderbans, contained a dominant proportion of Scheduled Caste in their population. Admittedly, zonal average of these villages pointed to the same direction of dominance of the Scheduled castes. But the data at the same time also hinted participation by local people, other than the Scheduled Castes. Cases in point were 6 villages of which 4 belonged to Zone I and 2 to Zone II.

Did these data indicate that (a) both the Scheduled and other castes, living within a certain range of proximity from the forests, were probably utilising the forest produce more and are likely to do so in the future also and (b) predominance of Scheduled Castes in far away villages, sending large number of groups, possibly pointed to the specialization of forest exploiting occupations by them? A question that evolved here was whether mastering of forest crafts was the result of eco-conditioning of villagers of nearby villages, irrespective of community identification. Alternatively, irrespective of distances, certain castes/ groups possessed and/or have acquired a fair degree of skill culturally that enabled them to carry on exploitation of forest in their struggle for survival. In other words, the roles they played were culture-specific, shaped by the ecological conditions.

# 4

# The Roles and the Risks

So far we have been analysing our data restricted to the permit holders of 157 villages. And one of the major ideas behind such examination was to get clues to understand the (i) nature and spread of activities concerned with forests; (ii) degree of involvement of the villages; and (iii) distinctive communities.

We now propose to supplement the above findings with certain other data focussing on different dimensions of social ecology. It appeared that these people or permit holders entering the forests were taking certain risks that could be broadly dichotomised as economical and physical. An examination of the nature and degree of these risks vis-a-vis the purposes of visit to the forests by the permit holders revealed interesting information. As obtained from the record of several field stations of the Project Tiger, the Government earned revenues of Rs. 56071.67, Rs. 236572.46 and Rs. 97545.50 against issuing the permits for fishing, timber and honey collection respectively, and also by selling one product itself during the three years of 1980-82 (for details see Table).

In terms of risks, the fishermen bear maximum economic risk with moderate degree of physical risk, for they usually take conditional loan from *Mahajan* or local money lenders. The conditions generally are to share 25 percent of the total catch with the *Mahajan*, over and above the interest levied on the total amount of the loan. The rate of interest varies from 15 to 30 percent according to the reputation of the group and their skill in the trade. The fishermen thus run a fair amount of economic risks in terms of losing their domestic materials, equipment etc., if they are unable to repay the loans due to failure in catch. They are however not greatly exposed to physical risks from man-eaters provided they avoid core areas and remain watchful and alert. But the tragedy is that the fish-population is quite low in the buffer zones, and significantly high in core areas. So the fishermen, drowned to the hair in debt, cannot always afford to avoid the core areas.

According to knowledgeable sources, the reason for such high and low population of fish is that urban effluent has been polluting the upland water-ways more than those of core areas situated close to the sea. Are the Sunderbans fishermen paying the price for the industrial development of Calcutta and Howrah with their lives?

The people who do not seem to run either the economic or physical risks or perhaps run a minimum of both, are the lumbermen or timber collectors. In most of the cases, they are employed persons and visit the jungles in large groups and with proper protection. Unfortunately, data in this score could not be collected during our enquiry. The above information was obtained in course of conversation with a number of people engaged in the trade.

A word now about the last occupation—collection of honey. As mentioned earlier, the Government earned substantial revenue by allowing the collectors to fetch honey from the forest. Honey thus collected cannot be sold by any private individual. It is purchased entirely by the Government who also markets it after de-waxing, if needed, and processing. In this total work activity, the honey-collectors' economic risks is minimum, if at all, since they are financed by the local bank. The Government, when purchasing the honey from the collectors, deduct the loan and pay up the remaining balance. Contextually, we must recall that while fishing is a perennial activity, honey is seasonal and is collected between March to June only.

The collectors of honey who practically do not bear any economic risk as such, run the maximum physical risk. They are required to collect honey from the forest blocks, most of which are situated either in the core areas and/or primitive areas. This is due to the fact that honey in this area is produced by the rock-bees (apis dorsata) which seldom build honeycombs in thinner forest blocks. A relatively higher death-rate among all the victims of man-eaters naturally come from these groups of honey-collectors.

Answers to the question why do they take such risks are: (a) it is culture-specific in the sense that the honey-collectors have traditionally learned this activity; (b) per capita income varies from Rs. 2000/- to Rs. 2200/- against working for 4 months which cannot be obtained from any other sources in the Sunderbans; (c) they can even supplement this income by earning through petty jobs, if and when available, for the rest of the years and (d) if any accident happens (which do occur frequently) their families will be entitled to compensation as sanctioned by the Government.

From the above information, an ecologically interesting point emerges. In the eco-setting of the Sunderbans, and with special reference to fishing and collection of honey, there is an inverse relationship between the two risks, namely, economic and physical. When one is high, the other is low and vice versa. And this inverse relationship allows these occupations to be pursued by the people. Obviously, they will not follow them if and when both the risks become high at once. The other ecologically important point is that while the peasants of the Sunderbans also run certain economic risks by borrowing and thereby being liable, at the most, to loose his land, these fishermen and honey collectors of the same area are taking enormous risk by staking their lives in pursuit for their livelihood.

Does this information indicate the magnitude of human sacrifice on the scaffold of socio-economic maladjustment of a backward area? We leave the question to the findings of surveys of future investigators.

Inferences, which we have tried so far to draw from our data are, more or less, of macro-level dimension. The following are the data of micro-level nature. In fact, it is a case-study of a particular Sunderban village. And we propose to close this chapter with the example of this village which is mostly utilising minor forest products for its survival. Hokaldanga* is a predominantly Muslim village in Mathurapore Police Station. It has about 500

---

*(Pseudonym)

Muslim families. There is a *madrassa* and a primary school but none of these institutions has much influence on the Muslim community. Apart from few families who pursue poultry-business, 27 families earn their livelihood by gathering and selling wax, honey and roots of wild trees. There are about 50 families whose adult members make a living as rag-pickers in Calcutta. There are also 20 families of beggars, besides a handful of wage labourers.

Of the total Muslim families, 80 percent live on leaves. It is only that instead of "two leaves and a bud", they survive on "there leaves". These are the leaves of *narkel* (coconut), *khejur* (date-palm) and *pichha.* They prepare *jhanta* (broom-stick), *knuste* (sweep-stick) and *pati* (mat) respectively from these leaves. The price of broom-stick varies from Rs. 6/- to Rs.14/- per dozen depending on whether the binding are made up of single or double-wires as also on their sizes. The price of *knuste* ranges between Rs. 20/- and Rs. 30/- per fifty according to the size and smoothness of the leaves.

In most cases, the manufacturers of these articles peddle their wares themselves at the following places: Kalighat, Chetla, Belghoria, Agarpara, Khardah, Titagarh, Kankinara, Ranaghat, Bandel, Tribeni, Barasat, Chakdah, and some other places.

It is interesting to note that they have chosen the Subarban areas as markets for their commodities, instead of the city, which would have been relatively easily accessible to them. Their arguments (needs to be probed in future) are that (a) the city people, like more fashionable and finished products and (b) they find it very difficult to peddle in the streets of the city which puzzle them with their complexities.

Like the poor of all places, they depend on borrowed money from the *mahajan* or money-lender for procuring raw materials and other necessities. In most cases, as we could gather, the rate of interest on such loan is Rs. 10 per day for a sum of Rs. 50 i.e., 720 percent per year.

In spite of their best efforts to collect these leaves from jungles at the risk of their lives, the irony of fate is that even then these leaves allow them food only for 9 months from *Bhadra* (September) to *Baisakh* (April-May). For the rest of the year they know not what or whom to depend on for food. But they do not complain. They are too poor to do so.

# 5

# The Dramas

Scenes change rather rapidly in this part of the delta. And most of the dramas are one-act plays.

Two big bulging eyes, set on a warty body, were watching and waiting. Suddenly, the moment arrived. The pair of powerful hind legs provided the adequate leverage for the action. Before it could realise what was happening, the body of the poor praying mantis was snapped by half under the powerful jaw. The frog got his food. It then hopped into the bush and headed for its resting place, searching on the way other insects to appease its ever hungry stomach.

It was the tide time. The boat drifted silently through the Neti Dhopani canal and stopped at a particular point of the island-shore. Four armed men alighted. They seemed to know their destination and were keen to achieve their purpose. It appeared that the topography of the land was at the fingertips of these poachers.

The tigress was guarding her week-old cubs while basking in the slanting rays of the setting sun. A moment earlier she quenched her thirst from a waterhole near the coastline. Her pug marks were a telltale evidence of her magnificent presence. She was tired and exhausted by the constant demand of her cubs for milk and by the effort of procuring her own food. Still she was alert enough to pick up any unfamiliar sound in the immediate vicinity. Suddenly she jumped up and some how could save her life. Silence was shattered by the booming of a gun. The bullet, however, missed the target. Instead, it struck her left hind paw. She did not hesitate, took a semi-circular turn and ran for the big river. More bullets followed. But the luck was with her this day. While running, she stumbled and the bullet aimed at hitting the base of her neck could pierce only a hole in one of her ears. She plunged into the river and swam across to reach the opposite bank. Her cubs were left behind to their fate. She could only carry with her the memory of biped animals with lightning stick.

On reaching the opposite bank she dragged her body, practically, on three legs towards a hental bush. Suddenly she frozed on her track. Few feets from her the hood of death rose. She had no intention to stretch her luck a bit any more. Moment later the snake lowered its hood and slighed into a nearby shrub. Sapped of all her energy she layed panting under the cover of the bush, unable even to lick her wounds.

The frog reached almost its destination. Gleefully it jumped, a little higher than normal, towards the pool.

But a pair of powerful beaks caught its poor body in the mid-air. The painted stork swallowed it in a gulp and flew towards its nest in the *bine* trees.

The black, sleek and glistning body slowly climbed the tree and reached the nest. A pair of small, beady and staring eyes surveyed silently the scene. The fantastically stretchable mouth swallowed one and then the second egg. But suddently the cacophany of birds sounded alarm calls and being charged by a group the snake had to forego the rest of the dinner and drop into the bush below. The painted stork lost two of its clutches of eggs.

Two eggs, at a time, was a bit too much for the snake. The free movement being hampered, it decided to rest and relax in the shade of a bush. But that was not to be. It suddenly sensed danger through the vibration in the ground. Soon it saw the approach of a *Tarkel* but was too slow to evade the attack or to run away.Was it repentent for its greed for the second egg? The powerful forelegs and the jaws of the monitor lizard *finished* the snake in a trice.

Nature ticked her ecological balance sheet where the debit-credit entries at the end of the day, always neutralise.

Leaning on the wings of the nocturnal birds, night descends slowly over this deltaic forest. Up above, the stars appear one by one. Their lights begin to reflect on the tidal canals below, surrounded by the green walls of the mangrove forests. Silence becomes all pervasive, and darkness steadily engulfs everything around.

It was the hour of false-dawn. Befooled by the hazy light, several solitary birds started chirping and then fell silent, as if ashamed of their mistake. Smoke-like mist was rising from the grayish surface of Matla. By and by the sun peeped through the foggy eastern horizon and shafts of tangent light plunged the world into a vermillion hue. Riverside villages were slowly shedding their lethargy to throb back into life. We cruised on towards more south and entered the core or primitive areas of the forest division.

River-Terns that were following us quietly withdrew. Even the sandpipers and snipes were reluctant to enter this area. The silence was nerve-wrecking and ocean-deep. One felt one could touch this wall of silence but refrained for fear of breaking it to invite a disaster.

Yes, disaster was the word that struck Jammuna Bibi when she learned that on and from the 14th of July she became a widow. Maqbul Sardar was a fisherman by profession. He had been fishing since his early boyhood days. The numerous kharis of Sunderbans were his professional play grounds both in the prime of his life and thereafter. He accompanied his father and kin members in all their fishing expeditions when he was young. Now, he was being accompanied by two of his nephews, one of whom replaced his son who was ill at home.

Two thoughts were tormenting Maqbul. So far the catch was not good enough. How could he repay at least some of the loan to the mahajan? This prompted him to take the risk of entering the core area. Seeing the fear written large on the faces of his nephews, he took up a spear and boasted that no living creatures dare to cross Maqbul's path and hope to live. With these encouraging words, he piloted the small boat into the narrower khari of Koikhali.

The second thought was more disturbing. When he left, he saw his son tossing on the bed with high fever and complaints of acute pain in the abdomen. Nothing he could do for him. Hospitalization was out of question since there was none within a rowing time of eight hours. He had left him to his fate and in the hands of older folks and their experiences.

Unmindfully Maqbul dropped down the spear by his side when an yellow streak of light leaped from behind the treacherously placid green mangrove forest and struck Maqbul directly. The tigress, as if flying in the air, carried the body of Maqbul in one swift movement on the other side.

For a few moments, oars in hands, the nephews sat like dumb dolls. Then they started shouting and moved towards the bank. The shattering roar reciprocated their shouts. It was a reply that no nuisance on any body's part will be tolerated. Jolted and terrified by the mighty roar they hurriedly turned the boat and rowed for their lives and for home. They reached home at a remarkably short time, only after 10 hours of hard rowing.

The tigress avenged the death of her cubs and injury to her leg. And Nature again crossed her balance sheet with a tick mark.

I met Maqbul's son this winter near the Naubeki water-ways in the Chamta Block. I asked why he had taken the risk of coming into the core area. The two big eyes stared at me seemingly to say what a foolish question. But he simply said that fish was plentiful in the area. His widow mother, two sisters and a little brother and, of course, his loving wife are all depending on him and his ability of catching fishes. Since such catches will be shared between his partners and that too after allowing a lion share to the *mahajan*, he had to take the risk. Besides, they do not know any alternative to fishing to maintain their families. When reminded of the earlier incident he paused and added as an after thought, "Shaheb, you earn from Government and pay taxes. We earn from the zemindary of Bon-Bibi and some times we are required to pay taxes. And if we abandon fishing in this area then there will be nothing left to us but starvation". He seemed to tremble as soon as the word `starvation' spilled out of his mouth. Was he mentally seeing a bunch of rickettelimbed, pot-bellied children, too weak to cry even, looking with glazed eyes towards the faces of their parents?

Deaths by starvation have recently been reported. Bhagyadhar Midda (55) of Kamdeb Nagar, Guinram Bhuyian (52), Giribala Bhuyian (70), Biswasundari Jana (38), Prafulla Mistri (35) and Champaki Sashmal (52) of Fatikpur have died of starvation. Footsteps of famine are sounding (Ananda Bazar Patrika—25.12.1982).

Dear reader, got the message? There is no escape for Maqbul's son and, for that matter, the sons of all Maqbuls. Hunger is million times more powerful and dangerous than maneaters. With some caution one however can avoid maneaters but the other peril is an all-destructive phenomenon here.

After a hard day's vigilance, the tired sun was setting. With four comrades, Maqbul's son was rowing towards Neti-Dhopanis canal, leaving behind long shadows on the water and ultimately disappeared under the foliage of garan trees. The rays of the setting sun turned the sky crimson. The river-terns returned and we turned for our homeward journey in the glowing evening sky of the Sunderbans, leaving only mute prayers for the well-being of Maqbul's son and his comrades.

# 6

# In Lieu of a Conclusion

The study of the ecosystem of the Sunderbans is a formidable and a challenging task. The biotic world of the delta is moving through two life-processes. One is illustrated by the network of mutual cooperation, leading sometime to symbiotic relationships e.g., that between deers and drongos, crocodiles and plovers, monkeys and deers etc. The other is the relationship of oppression and the waves of exploitation of the biotic and/or natural resources by man--the apex of the evolutionary pyramid.

The forests cannot avoid the axe of the lumberman. The fishes fail to escape the dragnet. The bees are smoked away. Birds are hunted for the family hearth. Snakes are captured and even the tigers are killed for pleasure. Nobody, I repeat, nobody can escape the meaningless slaughter, not even man by his own species.

Economic and social exploitation apart, cases of dacoity, arson, looting and murder are not infrequent.

Records of crime of the Sunderbans and facsimiles of clippings, from news papers as incorporated, tell their own stories. These data, however, represent only a tiny part of the total savagery.

Law and order situations, in the northern part of the Sunderbans, are controlled by five police stations. These are Hingalgunj, Sandeshkhali, Hasanabad, Haroa and Minakhan. Unfortunately the data on crime, as supplied, contained no information about the last named Police Station. Also data available were only for three years. Classificatory content analysis shows that these data can be arranged into a rank order. And what we find is a pattern of ranking of different types of crime committed in the three years between 1986 and 1988. It is seen that "other crime" than murder, dacoity, rioting, robbery, burglary and theft, occupied the top rank. It is followed by theft and rioting as second and third in rank.

Fluctuations, both by number and year of crimes commited, have also been noted in the data. For instance, slight decrease in the total number of recorded crime is observable in 1987 which again increased in 1988. Further while "rioting" occupied second position in 1986, encouragingly enough, it became relegated to third position in 1988.

In order to probe whether the above picture also holds true by Police Stations, data were analysed accordingly. Analysis shows that overall trend remaining constant, certain variations are located in respective Police Stations. For example, by pooling data of three years together, it appears that the area under Sandeshkhali Police Stations is relatively more "riot-prone" followed by

Hasanabad, and Haroa. Contextually, Hingalgunj seems to be a comparatively peaceful zone.

Till now, the crime situations of the northern part of the Sunderbans have been dealt with. Let us now peep into those situations then prevailing at the southern part of the Sunderbans.

Analyses of data of 10 Police Stations for 5 years, show a pattern somewhat different from their counterpart of the northern sector. Pooled data of the total period of reference exposed that; like the northern part, fluctuations both by number and year are also observable. Such fluctuations in terms of reference periods can be symbolically represented as:

1984> 1987> 1986> 1985> 1988. This indicates the decreasing degree of reported crimes between 1984 and 1988. Similarly, the overall patten of the nature of crime may be presented, in terms of their prevalences, as, Th> Md> Bg> Dc> Rb> (Th =Theft; Md = Murder; Bg = Burglary; Dc = Dacoity; Rb = Robbery), meaning 'theft' occupies the top rank and 'robbery' the lowest in the crime ladder of south Sunderbans. Against this general pattern, computation of respective crime by each year reveals another kind of variations. These are noted as:

Th=> 1984> 1986> 1985> 1987> 1988

Md=> 1987> 1985> 1986> 1988> 1984

Bg=> 1984> 1987> 1985> 1986> 1988

Dc=> 1987> 1984> 1986> 1988> 1985

Rb=> 1984> 1987> 1988> 1985> 1986

Apart from the year wise fluctuations, above data also indicate a matrix of crime which is dynamic or ever changing. As for example, while 'theft', 'burglary' and 'robbery' were "crimes of the year" in 1984, 'murder' and 'dacoity' were most prevalent during 1987. Considering only the first two positions in terms of number of crimes committed, it appears that the four years between, 1984 and 1987 were the blackest years when the southern Sunderbans were tortured under maximum number of dark deeds.

Let us now compare the crime situation between the northern and southern sectors of the Sunderbans. Comparable data available relate to the three years of 1986, 1987 and 1988. Also, crime that can be compared are 'murder', 'dacoity', 'robbery', 'burglary' and 'theft' being commonly recorded in both the parts of the Sunderbans during the above period of reference. What is observable through analysis is that number of cases against each of the crime is much more in the southern than in the northern part. For instance, in the southern part of the Sunderbans, 'murder', 'dacoity' and 'theft' occurred 3 times more than that of the northern part. But very interestingly no cases, either of 'rioting' or 'other crime' has been recorded in the south sector. This supports our investigative findings that persons who are fighting for their survival, of necessity, must cultivate integration among themselves. That they are doing so is amply expressed through participation in worshipping "Bon-bibi", "Dakhin Roy" and other deities by members of all community, irrespective of their religion, caste or creed.

Contextually, it is to be noted that the records of crime supplied by the authority is only a tip of the iceberg. In the Sunderbans it is a tremendously difficult proposition for the police stations responsible to control crimes. They are practically helpless where non-reporting of crime is the rule and exceptions are few and far between. Further, the intricate and difficult eco-condition of the delta acts like an impenetrable baffle wall to the forces of law and order which are very much limited in strength.

What I tried to convey here is that the inhabitants of the Sunderbans are passing through a continuous crisis of survivality. Even if by luck, one escapes from the Jaws of crocks and man eaters or snake bites, there is no guarantee that such a person will be able to enjoy his earnings from the forests peacefully later on. Bi-ped sharks and man-eaters are eagerly waiting by the side of the traps. Mouths of which remain open always.

Ironically, these bi-ped animals that created such vicious atmosphere parade this globe as saviour of environment, society and culture. They proudly claim top position on the evolutionary ladder. Paradoxically enough, some of them do fight death and destruction, corruption and pollution of both natural and social origin.

Possibly, it is the cell biology of man that are carrying good and bad elements simultaneously since its genesis. And although claimed to be a rational animal, more often than not he becomes a victim of irrationality and is much more attracted by dark deeds than otherwise. The bewildered earth is watching with an abeted breath the results of this relentless battle between Dr. Jackyle and Mr. Hyde among men.

This struggle is much more acute in this part of the earth. This whole deltaic region is gasping under the darkest cloud of exploitation, torture and sufferings. The greatest exploiter of the biotic world of this delta is Man. And the fiercest enemy is Hunger.

Dear reader, if this sketchy account has been able to communicate to you an iota of the evergoing battle on the eco-social existence of this delta my purpose has been served. One of the biggest among the mangrove deltas of the World craves for your support in the desperate bid for survival. Whether you will lend if or not the decision is yours.

## Acknowledgement

Summation of inputs like formulation of the theme of research, planning of enquiries, conducting surveys, collation and collection of field data as also execution of many vexing but none-the-less crucial field matters, have resulted in the output of this book.

I am thankful to so many persons for helping and guiding me in those said areas that to acknowledge each of them individually will cover more pages than the text itself. Naturally, the list of persons noted hereunder is by no means an exhaustive one. Under this situation any omission made inadvertantly is regretted. Such omission only indicates the level of inactivity of the "grey cells" of the author's brain.

I like to thank, first of all, the Indian Statistical Institute and its ex-Director, Professor (Dr.) B.P. Adhikari for allowing me to pursue this piece of research. Financial support was received from the Section 8(1) Committee of the Planning Commission to whom the acknowledgement is offered herewith.

I do not wish to undermine our friendship by offering thanks to one of my best friends, Professor (Dr.) Suraj

Bandyopadhyay for guiding in picking up and applying appropriate statistical tests as and where necessary.

I owe my thanks to Sri P. Sanyal, the then Director of the Sunderban Tiger Project, Sri Arun Chakraborty, former Range Officer and Sri Gopal Tanti of the same department for their cooperation and help.

To Professor (Dr.) Bikash Sinha for his counsel and to Sri Aniruddha Chakraborty, both of Indian Statistical Institute, for actively helping in collection, scrutiny and tabulation of field data. Sm. Swati Sadhu helped me much in computing the statistical tests. I thank her.

I am grateful to Late Ajoy Home, the well known naturalist for enhancing my knowledge on the ornithological aspects of the Sunderbans. I am equally indebted to Professor (Dr.) A.B. Gupta, former Dean of Studies, Indian Statistical Institute for his critical study, valuable suggestions and help in editing the manuscript. And that made it more readable.

I shall be failing in my duty if I do not acknowledge the invaluable secretarial assistances rendered by all the staff members of Sociological Research Unit, Indian Statistical Institute, specially Sarvasri Asish Kumar Basu and Narayan Ch. Saha.

Last, but not the least, I am grateful to all the villagers and interviewees who very patiently tolerated the encroachment on their time and privacy. The way they taught me how to study the problem has yet to be found in any text book of any University.

And I salute you, my dear reader, for your immense patience and indomitable spirit to read till this end.

**Table I**

**Distribution of villages in groups of Police Stations by different purpose for which forest permits were obtained**

| Police Stations | Honey (H) | | | | Fish (F) | | Timber | |
|---|---|---|---|---|---|---|---|---|
| | Only | With F | With F,T | With T | Only | With T | (T) only | Total |
| (1) | (2) | (3) | (4) | (5) | (6) | (7) | (8) | (9) |
| Gosaba, Basanti, Hingalganj, Sandeshkhali and Kultali | 10 | 11 | 11 | 4 | 20 | 6 | 12 | 74 |
| Hasnabad, Minakha, Canning and Mathurapur | - | - | 2 | - | 27 | 1 | 2 | 32 |
| Sonarpur, Bhangar, Bashirhat, | | | | | | | | |

(Contd.)

| | | | | | | | | |
|---|---|---|---|---|---|---|---|---|
| Pathorprotima, Kakdwip, Sagore, Kuldip, Falta, Diamond Harbour, and Budge Budge | - | - | - | - | 9 | 4 | 38 | 51 |
| Total | 10 | 11 | 13 | 4 | 56 | 11 | 52 | 157 |

**Table II**

**Frequency distribution of total number of groups with forest permit by number of villages**

| *Sr. No.* | *Total number of groups* | *Number of villages* | *Sr. No.* | *Total number of groups* | *Number of villages* |
|---|---|---|---|---|---|
| (1) | (2) | (3) | (1) | (2) | (3) |
| 1 | 1 | 36 | 23 | 27 | 1 |
| 2 | 2 | 13 | 24 | 30 | 1 |
| 3 | 3 | 9 | 25 | 31 | 1 |
| 4 | 4 | 5 | 26 | 37 | 1 |
| 5 | 5 | 9 | 27 | 39 | 1 |
| 6 | 6 | 10 | 28 | 40 | 1 |
| 7 | 7 | 3 | 29 | 45 | 2 |
| 8 | 8 | 4 | 30 | 52 | 1 |
| 9 | 9 | 3 | 31 | 58 | 1 |
| 10 | 10 | 2 | 32 | 62 | 1 |
| 11 | 11 | 3 | 33 | 64 | 1 |
| 12 | 12 | 4 | 34 | 73 | 1 |
| 13 | 14 | 2 | 35 | 86 | 1 |
| 14 | 15 | 3 | 36 | 88 | 1 |
| 15 | 16 | 1 | 37 | 104 | 1 |
| 16 | 17 | 1 | 38 | 114 | 1 |
| 17 | 20 | 3 | 39 | 125 | 1 |
| 18 | 22 | 1 | 40 | 130 | 1 |
| 19 | 23 | 6 | 41 | 146 | 1 |
| 20 | 24 | 2 | 42 | 153 | 1 |
| 21 | 25 | 1 | 43 | 154 | 1 |
| 22 | 26 | 2 | 44 | 158 | 1 |

(Contd.)

| | | |
|---|---|---|
| 45 | 172 | 1 |
| 46 | 176 | 1 |
| 47 | 177 | 1 |
| 48 | 188 | 1 |
| 49 | 193 | 1 |
| 50 | 267 | 1 |
| 51 | 284 | 1 |
| 52 | 319 | 1 |
| 53 | 388 | 1 |
| 54 | 458 | 1 |
| 55 | 1154 | 1 |
| Total | 6483 | 157 |

**Table III**

**Total number of groups with forest permits, their average number per village and percent of villagesout of total inhabited villages in a Police Station**

| *Zone* | *Sr. No.* | *Police Station* | *Total no. of groups* | *Average no. of groups per village* | *percentage of 'permit holder villages" out of total inhabited villages* |
|---|---|---|---|---|---|
| (1) | | (2) | (3) | (4) | (5) (6) |
| Within "Sunderban" area* | | | | | |
| Nearest | 1 | Gosaba | 3147 | 104.90 | 29.13 |
| | 2 | Hingalganj | 847 | 77.00 | 35.48 |
| | 3 | Sandeshkhali | 83 | 7.55 | 19.64 |
| | 4 | Basanti | 648 | 43.20 | 24.19 |
| | 5 | Kultali | 36 | 5.14 | 14.00 |
| Nearer | 6 | Hasnabad | 315 | 35.00 | 7.76 |
| | 7 | Minakhan | 52 | 17.33 | 4.69 |
| | 8 | Canning | 505 | 36.07 | 7.45 |
| | 9 | Mathurapur | 167 | 27.83 | 4.38 |

* Administrative area of Sunderban

| | | | | | |
|---|---|---|---|---|---|
| Distant | 10 | Latharprotima | 112 | 7.00 | 18.39 |
| | 11 | Kakdwip | 12 | 3.00 | 8.33 |
| | 12 | Sagore | 1 | 1.00 | 4.17 |
| | 13 | Namkhana | - | - | - |
| | 14 | Jaynagar | - | - | - |
| | 15 | Haroa | - | - | - |
| Outside "Sunderban" area | | | | | |
| | 16 | Bashirhat | 47 | 7.83 | 4.65 |
| | 17 | Bhangar | 120 | 60.00 | 1.40 |
| | 18 | Sonarpur | 202 | 67.33 | 3.06 |
| | 19 | Kulpi | 148 | 12.33 | 4.86 |
| | 20 | Diamond Harbour | 28 | 7.00 | 2.40 |
| | 21 | Falta | 2 | 1.00 | 1.54 |
| | 22 | Budge Budge | 11 | 11.00 | 1.09 |
| | | Total | 6483 | 531.51 | 196.61 |
| | | Contiguity ratio | | 0.62+0.2325 ($0.045<p<0.05$) | 0.51+0.2325 ($0.015<p<0.02$) |

**Table IV**

**Some characteristics of frequency distributions of number of groups of permit holders from a village***

| *Characteristics* | *Nearest* | *Nearer* | *Distant* | *Total* |
|---|---|---|---|---|
| (1) | (2) | (3) | (4) | (5) |
| Mode | 1 | 1 | 1 | 1 |
| Median | 8 | 6.50 | 6 | 6 |

(Contd.)

| | | | | |
|---|---|---|---|---|
| Mean | 64.34 | 32.47 | 13.39 | 41.29 |
| Standard Deviation | 157.99 | 49.75 | 28.74 | 114.27 |
| Coefficient of variation | 245.55 | 153.22 | 214.64 | 276.75 |
| (N) | (74) | (32) | (51) | (157) |

* Nearest to the forests = Zone I
Nearer to the forests = Zone II
Distant from the forests = Zone III

**Table V**

**Number of villages in a Police Station in each zone from which groups of permit holders come**

| *No.of groups* | *Number of villages* | | | *No. of groups* | *Number of villages* | | |
|---|---|---|---|---|---|---|---|
| | *Zone 1* | *Zone 2* | *Zone 3* | | *Zone 1* | *Zone 2* | *Zone 3* |
| (1) | (2) | (3) | (4) | (1) | (2) | (3) | (4) |
| 1 | 15 | 6 | 15 | 11 | - | 1 | 2 |
| 2 | 6 | 4 | 3 | 12 | 2 | - | 2 |
| 3 | 2 | 3 | 4 | 14 | 1 | - | 1 |
| 4 | 3 | 2 | - | 15 | 1 | 1 | 1 |
| 5 | 6 | - | 3 | 16 | - | - | 1 |
| 6 | 2 | 1 | 7 | 17 | - | - | 1 |
| 7 | 2 | 1 | - | 20 | 2 | - | 1 |
| 8 | 2 | - | 2 | 22 | 1 | - | - |
| 9 | 1 | 1 | 1 | 23 | 3 | 2 | 1 |
| 10 | 2 | - | - | 24 | 2 | - | - |

(Contd.)

| | | | |
|---|---|---|---|
| 25 | - | - | 1 |
| 26 | 1 | - | 1 |
| 27 | 1 | - | - |
| 30 | - | 1 | - |
| 31 | - | - | 1 |
| 37 | 1 | - | - |
| 39 | 1 | - | - |
| 40 | - | 1 | - |
| 45 | - | 1 | 1 |
| 52 | 1 | - | - |
| 58 | 1 | - | - |
| 62 | - | 1 | - |
| 64 | 1 | - | - |
| 73 | - | 1 | - |
| 86 | - | 1 | - |
| 88 | - | 1 | - |
| 104 | 1 | - | - |
| 114 | - | - | 1 |
| 125 | - | 1 | - |
| 130 | 1 | - | - |
| 146 | 1 | - | - |
| 153 | 1 | - | - |
| 154 | 1 | - | - |
| 158 | 1 | - | - |
| 172 | - | 1 | - |
| 176 | - | - | 1 |
| 177 | 1 | - | - |
| 188 | 1 | - | - |
| 193 | - | 1 | - |
| 267 | 1 | - | - |
| 284 | 1 | - | - |
| 319 | 1 | - | - |
| 388 | 1 | - | - |
| 458 | 1 | - | - |
| 1154 | 1 | - | - |
| Total 16483 | 74 | 32 | 51 |

## Table VI

**Result of tests of significance between frequency distributions of total number of groups of permit holders from a village by different zones**

**(Kolmogorov-Smirnov's D-Statistics)**

| Zone | Sub-Zone | *Nearest (Zone I)* | | | | |
|---|---|---|---|---|---|---|
| | | *Gosaba* | *Hingalganj, Basanti, Sandeshkhali, Kultali* | *Combined* | *Nearer (Zone I)* | *Distant (ZoneII)* |
| (1) | (2) | (3) | (4) | (5) | (6) | (7) |
| Nearest (Zone I) | Gosaba | - | 0.93 (0.10<P) | | | |
| | Hingalganj, Basanti, Sandeshkhali, Kultali | | | | | |
| | Combined | | | | 2.50 (P<0.001) | 3.11 (P<0.001) |
| Near (Zone II) | - | | | | - | 0.39 (p>0.10) |
| Distant (Zone III) | - | | | | | - |

## Table VII

**Number of groups and villages involved in forest permits**

| Zone | No. of groups with forest permits | | | | No. of villages involved | | | | Total no. of inhabited |
|---|---|---|---|---|---|---|---|---|---|
| | Fish | Honey | Timber | Total | Fish | Honey | Timber | Total | villages in a police station |
| (1) | (2) | (3) | (4) | (5) | (6) | (7) | (8) | (9) | (10) |
| Nearest | 3802 | 560 | 399 | 4761 | 48 | 36 | 33 | 74 | 302 |
| Nearer | 1020 | 9 | 10 | 1039 | 30 | 2 | 5 | 32 | 505 |
| Distant | 421 | - | 262 | 683 | 13 | - | 42 | 51 | 1468 |
| Total | 5243 | 569 | 671 | 6483 | 91 | 38 | 80 | 157 | 2275 |

**Table VIII**

**No. of groups with forest permits, no. of villages involved in each purpose both within and outside officially notified "Sundarban" area**

| Zone | Sr. No. | Police Station | Number of groups with forest permit | | | | Number of villages involved | | | | Total no. of inhabited villages in a Police Station |
|---|---|---|---|---|---|---|---|---|---|---|---|
| | | | Fish | Honey | Timber | Total | Fish | Honey | Timber | Total | |
| (1) | (2) | (3) | (4) | (5) | (6) | (7) | (8) | (9) | (10) | (11) | (12) |
| **Within "Sunderban" area* :** | | | | | | | | | | | |
| Nearest to the forest | 1 | Gosaba | 2548 | 326 | 273 | 3147 | 24 | 16 | 10 | 30 | 103 |
| | 2 | Hingalganj | 673 | 148 | 26 | 847 | 9 | 9 | 4 | 11 | 31 |
| | 3 | Sandeshkhali | 5 | 30 | 48 | 83 | 3 | 3 | 9 | 11 | 56 |
| | 4 | Basanti | 556 | 43 | 49 | 648 | 8 | 5 | 9 | 15 | 62 |
| | 5 | Kultali | 20 | 13 | 3 | 36 | 4 | 3 | 1 | 7 | 50 |

*Administrative area of sunderban ıderban

(Contd.)

| (1) | (2) | (3) | (4) | (5) | (6) | (7) | (8) | (9) | (10) | (11) | (12) |
|---|---|---|---|---|---|---|---|---|---|---|---|
| Nearer to the forest | 6 | Hasnabad | 309 | 1 | 5 | 315 | 9 | 1 | 1 | 9 | 116 |
| | 7 | Minakhan | 52 | - | - | 52 | 3 | - | - | 3 | 64 |
| | 8 | Canning | 501 | - | 4 | 505 | 12 | - | 3 | 14 | 188 |
| | 9 | Mathurapur | 158 | 8 | 1 | 167 | 6 | 1 | 1 | 6 | 137 |
| Distant from the forest | 10 | Pathorpratima | 5 | - | 107 | 112 | 1 | - | 16 | 16 | 87 |
| | 11 | Sagore | 1 | - | - | 1 | 1 | - | - | 1 | 24 |
| | 12 | Kakdwip | - | - | 12 | 12 | - | - | 4 | 4 | 48 |
| | 13 | Namkhana | - | - | - | - | - | - | - | - | 25 |
| | 14 | Joynagar | - | - | - | - | - | - | - | - | 139 |
| | 15 | Haroa | - | - | - | - | - | - | - | - | 139 |
| **Outside "Sunderban" area :** | | | | | | | | | | | |
| | 16 | Bashirhat | 3 | - | 44 | 47 | 2 | - | 6 | 6 | 129 |
| | 17 | Bhangar | 118 | - | 2 | 120 | 2 | - | 1 | 2 | 143 |

(Contd.)

| (1) | (2) | (3) | (4) | (5) | (6) | (7) | (8) | (9) | (10) | (11) | (12) |
|---|---|---|---|---|---|---|---|---|---|---|---|
| 18 | Sonarpur | | 202 | - | - | 202 | 3 | - | - | 3 | 98 |
| 19 | Kulpi | | 87 | - | 61 | 148 | 3 | - | 9 | 12 | 247 |
| 20 | Diamond Harbour | | 5 | - | 23 | 28 | 1 | - | 3 | 4 | 167 |
| 21 | Falta | | - | - | 2 | 2 | - | - | 2 | 2 | 130 |
| 22 | Budge Budge | | - | - | 11 | 11 | - | - | 1 | 1 | 92 |
| | Total | | 5243 | 569 | 671 | 6483 | 91 | 38 | 80 | 157 | 2275 |

## Table IX

**Police Station-wise average number of groups of permit holders and percentage of villages from which they come to total number of inhabited villages**

| Sr. No. | Police Station | *Average size of the group* | | | *Percentage of villages to total no. of inhabited in each P.S.* | | |
|---|---|---|---|---|---|---|---|
| | | *Fish* | *Honey* | *Timber* | *Fish* | *Honey* | *Timber* |
| (1) | (2). | (3) | (4) | (5) | (6) | (7) | (8) |
| 1 | Gosaba | 106.17 | 20.38 | 27.30 | 23.30 | 15.53 | 9.70 |
| 2 | Hingalganj | 74.78 | 16.44 | 6.50 | 29.03 | 29.03 | 12.90 |
| 3 | Sandeshkhali | 1.67 | 10.00 | 5.33 | 5.36 | 5.36 | 16.07 |
| 4 | Basanti | 69.50 | 8.60 | 5.44 | 12.92 | 8.06 | 14.52 |
| 5 | Kultali | 5.00 | 4.33 | 3.00 | 8.00 | 6.00 | 2.00 |
| 6 | Hasnabad | 34.33 | 1.00 | 5.00 | 7.76 | 0.86 | 0.86 |
| 7 | Minakhan | 17.33 | - | - | 4.69 | - | - |
| 8 | Canning | 41.75 | - | 1.33 | 6.38 | - | 1.60 |
| 9 | Mathurapur | 26.33 | 8.00 | 1.00 | 4.38 | 0.73 | 0.73 |

(Contd.)

| (1) | (2) | (3) | (4) | (5) | (6) | (7) | (8) |
|---|---|---|---|---|---|---|---|
| 10 | Bashirhat | 1.50 | - | 7.33 | 1.55 | - | 4.65 |
| 11 | Bhangar | 59.00 | - | 2.00 | 1.40 | - | 0.70 |
| 12 | Sonarpur | 67.33 | - | - | 3.06 | - | - |
| 13 | Pathorpratima | 5.00 | - | 6.69 | 1.15 | - | 8.33 |
| 14 | Kakdwip | - | - | 3.00 | - | - | 8.33 |
| 15 | Kulpi | 29.00 | - | 6.78 | 1.21 | - | 3.64 |
| 16 | Sagore | 1.00 | - | - | 4.17 | - | - |
| 17 | Diamond Harbour | 5.00 | - | 7.67 | 6.60 | - | .1.80 |
| 18 | Falta | - | - | 1.00 | - | - | 1.54 |
| 19 | Budge Budge | - | - | 11.00 | - | - | 1.09 |
| | Contiguity ratio (C) | 0.8240 | 0.4706 | 0.7979 | 0.5045 | 0.6127 | 0.9246 |
| | (C-1)/Sd(C) | -0.76 | -2.28 | -0.87 | -2.13 | -1.67 | -0.32 |
| | P | - | 0.010<P< 0.015 | - - | 0.015<P< 0.020 | 0.045<P< 0.050 | - |

## Table X

**Results of tests of significance between frequency distributions of groups of permit holders from a village by different purposes.**

**(Kolmogorov-Smiranov's D-Statistics)**

| *Purpose* | *Number of villages* | *Purpose* | | |
|---|---|---|---|---|
| | | *Fish* | *Honey* | *Timber* |
| (1) | (2) | (3) | (4) | (5) |
| Fish | 91 | - | 0.2430 | 0.3676 |
| | | | (0.05<P<0.10) | (P<0.01) |
| Honey | 38 | - | - | 0.2263 |
| | | | | (0.10<P) |
| Timber | 80 | - | - | - |

Table XI

Spatial distinctiveness of the area

| *Characteristics of a police station* | | | Obtained Spearman's rank correlation coefficient of each characteristic with total no. of groups of permit holders in a police station [N=19] |
|---|---|---|---|
| *Type* | *Sr. No.* | *Description* | |
| (1) | (2) | (3) | (4) |
| Demographic | 1 | Total male population | -0.0441 |
| | 2 | Population density | -0.2647 |
| Economic | 1 | Percentage unemployed among total male population | -0.0759 |
| | 2 | Percent engaged in agriculture out of total male population | 0.3089 |

(Contd.)

| (1) | (2) | (3) | (4) |
|---|---|---|---|
| | 3 | Percent working as agricultural labourer out of total male worker | 0.4461* |
| | 4 | Percent working as agriculture labourer out of total male engaged in agriculture | -0.1740 |
| | 5 | Total number of male cultivator | 0.3775 |
| | 6 | Total number of male agriculture labourers | 0.3334 |
| Community | 1 | Total scheduled caste male population | 0.6185** |
| | 2 | Total scheduled tribe male population | 0.6079** |
| | 3 | Total scheduled male population | 0.7280** |

* Significant at 10% level

** Significant at 1 % level

**Table XII**

**Percentage of scheduled castes (male) population by police stations in each zone arranged in descending order (number of villages involved shown within brackets)**

| | Zone | | |
|---|---|---|---|
| | Nearest | Nearer | Distant |
| | (1) | (2) | (3) |
| | 71.60 | 64.70 | 63.10 |
| | (11) | (14) | (2) |
| | 67.70 | 44.70 | 52.40 |
| | (30) | (6) | (3) |
| | 65.90 | 40.70 | 41.80 |
| | (11) | (9) | (2) |
| | 48.30 | 22.70 | 36.40 |
| | (7) | (3) | (12) |
| | | | 30.30 |
| | | | (4) |
| | | | 25.00 |
| | | | (6) |
| | | | 19.30 |
| | | | (4) |
| | | | 16.40 |
| | | | (1) |
| | | | 16.10 |
| | | | (16) |
| | | | 13.10 |
| | | | (1) |
| Percentage for each zone | 62.60<br>(74) | 43.20<br>(32) | 31.39<br>(51) |

**Table XIII**

**Percentage of scheduled castes in villages under each zone sending large number of groups to forest**

| Size of the groups | Zone | | |
|---|---|---|---|
| | Nearest | Nearer | Distant |
| (1) | (2) | (3) | (4) |
| 104 | 79.83 | - | - |
| 114 | - | - | 85.33 |
| 125 | - | 31.76 | - |
| 130 | 72.62 | - | - |
| 146 | 38.52 | - | - |
| 153 | 42.05 | - | - |
| 154 | 66.55 | - | - |
| 158 | 85.65 | - | - |
| 167 | 87.09 | - | - |
| 172 | - | 22.00 | - |
| 176 | - | - | 62.45 |
| 177 | 19.44 | - | - |
| 188 | 93.92 | - | - |
| 193 | - | 95.49 | - |
| 284 | 78.95 | - | - |
| 319 | 00.20 | - | - |
| 388 | 98.73 | - | - |
| 458 | 90.47 | - | - |
| 1154 | 63.00 | - | - |

## Table XIV

**Revenue collected in a year from the permit holders visiting Sunderban for different purposes**

| *Sr. No.* | *Purposes* | *Annual revenue (in Rs.)* 1980 | 1981 | 1982 (upto June) | *Total* |
|---|---|---|---|---|---|
| (1) | (2) | (3) | (4) | (5) | (6) |
| 1 | Fish | 20569.00 | 24567.00 | 10935.67 | 56071.67 |
| 2 | Honey | 24563.00 | 31299.50 | 41683.00 | 97545.50 |
| 3 | Timber | 1430.32 | 111856.34 | 123285.90 | 236572.56 |
| | Total | 46562.32 | 167722.84 | 175904.57 | 390189.73 |

## Table XV

**Distribution of villages by purposes in each number of groups**

| No. of groups | Villages by each purpose | | | No. of groups | Villages by each purpose | | |
|---|---|---|---|---|---|---|---|
| | Fish | Honey | Timber | | Fish | Honey | Timber |
| (1) | (2) | (3) | (4) | (1) | (2) | (3) | (4) |
| 1 | 14 | 8 | 29 | 8 | 1 | 5 | 3 |
| 2 | 12 | 2 | 8 | 9 | 2 | - | 3 |
| 3 | 4 | 3 | 8 | 10 | 1 | - | - |
| 4 | 5 | 1 | 2 | 11 | 2 | 1 | 2 |
| 5 | 3 | 3 | 4 | 12 | 1 | - | 3 |
| 6 | 4 | 1 | 5 | 14 | - | - | 2 |
| 7 | 2 | 2 | 1 | 15 | 2 | 1 | 1 |

(Contd.)

| | | | | | | | |
|---|---|---|---|---|---|---|---|
| 16 | - | - | 2 | 40 | 1 | - | - |
| 17 | - | 1 | 1 | 43 | - | 1 | - |
| 18 | - | - | 2 | 45 | 2 | - | - |
| 20 | 2 | 2 | - | 56 | - | 1 | - |
| 22 | - | 1 | - | 62 | 1 | - | - |
| 23 | 2 | 2 | - | 64 | 2 | - | - |
| 24 | 1 | - | 1 | 71 | - | - | 1 |
| 25 | 1 | - | - | 86 | 2 | - | - |
| 26 | 1 | - | 1 | 104 | 1 | - | - |
| 30 | 1 | 1 | - | 110 | 1 | - | - |
| 31 | 1 | - | - | 112 | 1 | - | - |
| 32 | 1 | - | - | 117 | 1 | - | - |
| 35 | 1 | - | - | 125 | 1 | - | - |
| 39 | - | 1 | - | 138 | 2 | - | - |

(Contd.)

| | | | | | | | |
|---|---|---|---|---|---|---|---|
| 139 | 1 | - | - | 226 | 1 | - | - |
| 150 | - | 1 | - | 264 | 1 | - | - |
| 176 | 2 | - | - | 264 | 1 | - | - |
| 179 | - | - | 1 | 344 | 1 | - | - |
| 188 | 1 | - | - | 379 | 1 | - | - |
| 193 | 1 | - | - | 825 | 1 | - | - |

**Table XVI**

**Some characteristics of distribution of groups of villagers going to the forest for different purposes**

| *Characteristics* | *Fishing* | *Honey* | *Timber* | *Total* |
|---|---|---|---|---|
| (1) | (2) | (3) | (4) | (5) |
| Modal number of group | 1 | 1 | 1 | 1 |
| Median | 9 | 7 | 3 | 6 |
| Average number of group per village | 57.62 | 14.97 | 8.39 | 41.29 |
| Range | 824 | 149 | 178 | 1153 |
| Standard Deviation | 114.32 | 25.63 | 21.32 | 114.70 |
| Coefficient of Variation | 198.40 | 171.21 | 254.11 | 276.75 |
| (N) | (91) | (38) | (80) | (157) |

Table XVII

**Distribution of permit holders by size class and purposes**

| *Size class by number of groups with forest permits from a village* | *Percentage distribution of villages by purpose of permit* | | | |
|---|---|---|---|---|
| | *Fish* | *Honey* | *Timber* | *Total* |
| (1) | (2) | (3) | (4) | (5) |
| 1-9 | 51.65 | 65.78 | 78.75 | 58.59 |
| 10-30 | 15.38 | 23.68 | 18.75 | 21.02 |
| Sub-total | 67.02 | 89.46 | 97.50 | 79.61 |
| 31-99 | 12.09 | 7.89 | 1.25 | 8.28 |
| 101 + | 20.88 | 2.65 | 1.25 | 12.11 |
| Sub-total | 32.97 | 10.54 | 2.50 | 20.39 |
| Total | 100.00 | 100.00 | 100.00 | 100.00 |
| n | 91 | 38 | 80 | 157 |

## SELECTED BIBLIOGRAPHY


Allan, B.C. (et al),

1979 - Gazetteer of Bengal and North-East India.

Ascoli, F.D.,

1921 - Revenue History of Sunderban (From 1870 to 1920).- Calcutta

Census 1951 - District 24-Parganas, Director of Census operations of West Bengal and Sikkim.

—— 1961 - Do

—— 1971 - Do

Chattopadhyay, K.

1989 - "From Mangrove to Temperate Forests in West Bengal" in B. Chaudhuri (ed), Forest and Forest Development in India, Inter-India Publications, New Delhi.

—— 1990 - Social Ecology of a Forest Delta, MAN IN INDIA, Vol. 70, No. 3, pp :

253-265.

Chaudhuri, A.B. and Chakrabarti, K.

1989 - Sunderbans Mangrove (Ecology and Wild Life).

Das, Amal Kumar
and

Raha, Manis Kumar

1963 - The Oraons of Sunderban Special series no. 3, Bulletin of the Cultural Research Institute, Calcutta.

Das, Amal Kumar (et al),

1981 - a focus on Sunderban Editions Indian, Calcutta.

Freilich, M.

1963 - "TheNationalexperiment, ecology and culture", South-Western Journal of Anthropology.

Heining, R.L.

1982 - Working-Plan of the Sunderban Government Forest in Khulna and 24-Parganas District, Bengal, Calcutta.

Hunter, W.W.,

1877 - A Statistical account of Bengal, Vol. 1, District of 24-Parganas and Sunderban. Delhi.

Knight Gregory, C.

1974 - Ecology and Change. New York: Academic Press.

Mukerjee, Radhakamal,

1940 - Man and His Habitation: A study in Social Ecology. Bombay, Popular Prakashan.

—— 1943 - Social Ecology. Longmans, Green & Co., Calcutta.

Mukherji, Shankarananda

1964 - "The Mau-leys and the Bau-leys of the Sunderban". Bulleting of the Cultural Research Institute, Vol.II, No. 2, pp. 56-58.

O'Malley, I.S.S.

1914 - Gazetteers of twenty four Parganas District. Bengal.

Pargiter, F.E.

1885 - Revenue History of the Sunderbans (From 1765 to 1870). Calcutta.

Sturte Vant, W.C.

1965 - "Studies in ethnoscience". American Anthropologist, 66: 3, part 2: 99-131.

The Sunderban Development Board

1976 - Sunderban: Promise of Growth Centres, West Bengal.

—— 1979 - Project Report on Sunderban Development, West Bengal.

United Bank of India,

1972 - The Report of the Fact Finding Survey on 24-Parganas District, West Bengal.

—— 1977 - Credit Plan for South 24-Parganas District, West Bengal.

